Bizarre Beast Battles
King Cobra vs. Bald Eagle
I0813892
Gareth Stevens
Publishing
By Natalie Humphrey

Please visit our website, www.garethstevens.com. For a free color catalog of all our high-quality books, call toll free 1-800-542-2595 or fax 1-877-542-2596.

Library of Congress Cataloging-in-Publication Data
Names: Humphrey, Natalie, author.
Title: King cobra vs. bald eagle / Natalie Humphrey.
Other titles: King cobra versus bald eagle
Description: New York : Gareth Stevens Publishing, [2023] | Series: Bizarre beast battles | Includes bibliographical references and index.
Identifiers: LCCN 2022024793 | ISBN 9781538284056 (library binding) | ISBN 9781538284032 (paperback) | ISBN 9781538284063 (ebook)
Subjects: LCSH: King cobra–Juvenile literature. | Bald eagle–Juvenile literature.
Classification: LCC QL666.O64 H866 2023 | DDC 597.96/42–dc23/eng/20220630
LC record available at https://lccn.loc.gov/2022024793

First Edition

Published in 2023 by
Gareth Stevens Publishing
2544 Clinton Street
Buffalo, NY 14224

Designer: Leslie Taylor
Editor: Natalie Humphrey

Photo credits: Cover (cobra) Roberto 33/Shutterstock.com; cover (eagle) KarSol/Shutterstock.com; series art (background texture) Apostrophe/Shutterstock.com; series art (eagle icon) masmascot/Shutterstock.com; series art (cobra icon) Credo Graphics/Shutterstock.com; pp. 4 & 6 Andrei Minsk/Shutterstock.com; pp. 5 & 7 Robert Enriquez/Shutterstock.com; p. 8 Skynavins/Shutterstock.com; p. 9 WildlifeBrian/Shutterstock.com; p. 10 Krit Leoniz/Shutterstock.com; p. 11 Alexandre Boudet/Shutterstock.com; p. 12 BIOSPHOTO/Alamy.com; p. 13 ChasingLightPhotos/Shutterstock.com; p. 14 Pornchaiwaiyakarn/Shutterstock.com; p. 15 TeusRenes/Shutterstock.com; p. 16 RealityImages/Shutterstock.com; p. 17 Christopher Zimmer/Alamy.com; p. 18 Suresh Suryasree/Shutterstock.com; p. 19 Jon C. Beverly/Shutterstock.com; p. 21 (cobra) AlinaMD/Shutterstock.com; p. 21 (eagle) Sergey Uryadnikov/Shutterstock.com.

Printed in the United States of America

CPSIA compliance information: Batch #CWGS23: For further information contact Gareth Stevens at 1-800-542-2595.

CONTENTS

Words in the glossary appear in **bold** type the first time they are used in the text.

KING OF SNAKES

Native to southern and Southeast Asia, the king cobra is known for its way of scaring off predators. When there is a **threat**, the king cobra lifts its body into the air and spreads out the back of its neck. This is meant to make the snake look bigger, but if that doesn't work, the king cobra has another trick. Striking with a **venom** deadly enough to kill an elephant, the king cobra is hard to beat.

But is that enough to help this snake **slither** to victory?

Explore Team

Dad makes
a haunted house.
There are bats
and spiders.

The king cobra is in the same family as the mamba, the coral snake, and the taipan.

RAPTORS IN THE SKY

Found only in North America, the bald eagle is a national **symbol** of the United States! This giant **raptor** is known for its white head, brown body, and giant wings.

While this eagle will eat nearly anything it can catch, it likes fish best and makes its home close to water. When bald eagles can't catch their food, they will steal from another bird. This hunter won't lose this battle without a fight!

NORTH AMERICA

BALD EAGLE RANGE

When hungry, bald eagles may also eat human garbage.

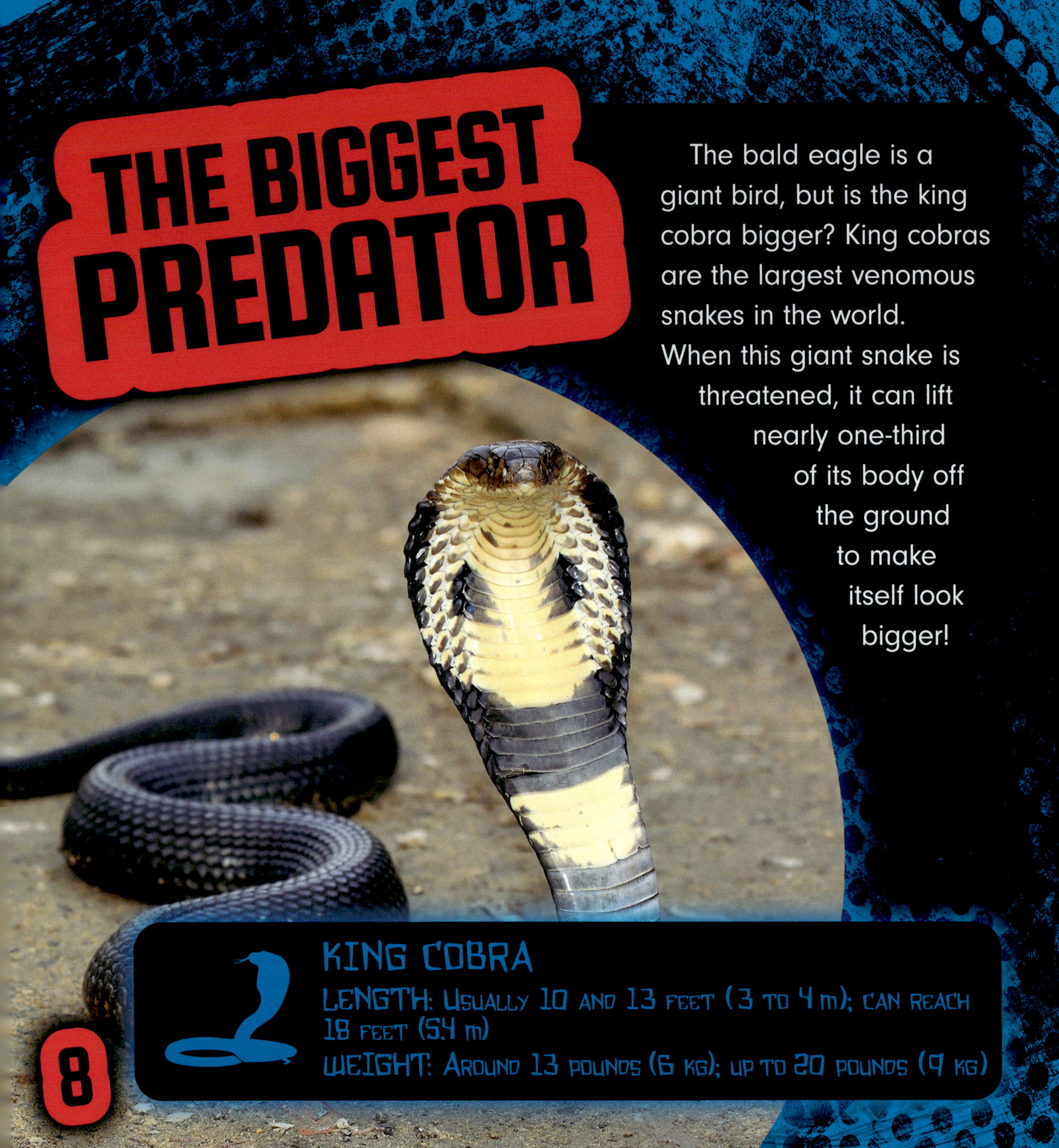

THE BIGGEST PREDATOR

The bald eagle is a giant bird, but is the king cobra bigger? King cobras are the largest venomous snakes in the world. When this giant snake is threatened, it can lift nearly one-third of its body off the ground to make itself look bigger!

KING COBRA

LENGTH: Usually 10 and 13 feet (3 to 4 m); can reach 18 feet (5.4 m)

WEIGHT: Around 13 pounds (6 kg); up to 20 pounds (9 kg)

BALD EAGLE

WINGSPAN: MALES: 6.6 FEET (2 M); FEMALES: 8 FEET (2.5 M)
LENGTH: MALES: 36 INCHES (90 CM); FEMALES: 43 INCHES (108 CM)

Bald eagles have a huge wingspan, or the length from the tip of one wing to the tip of the other. In fact, a female eagle's wingspan can be even wider than a human is tall. In this battle of sizes though, the king cobra has the bald eagle beat!

A DEADLY BITE AND GRIP

The king cobra is known for its **fangs** that deliver deadly venom. A venomous strike from a king cobra can kill a person in close to 30 minutes! The king cobra doesn't go after humans unless threatened, but this snake should be given lots of space!

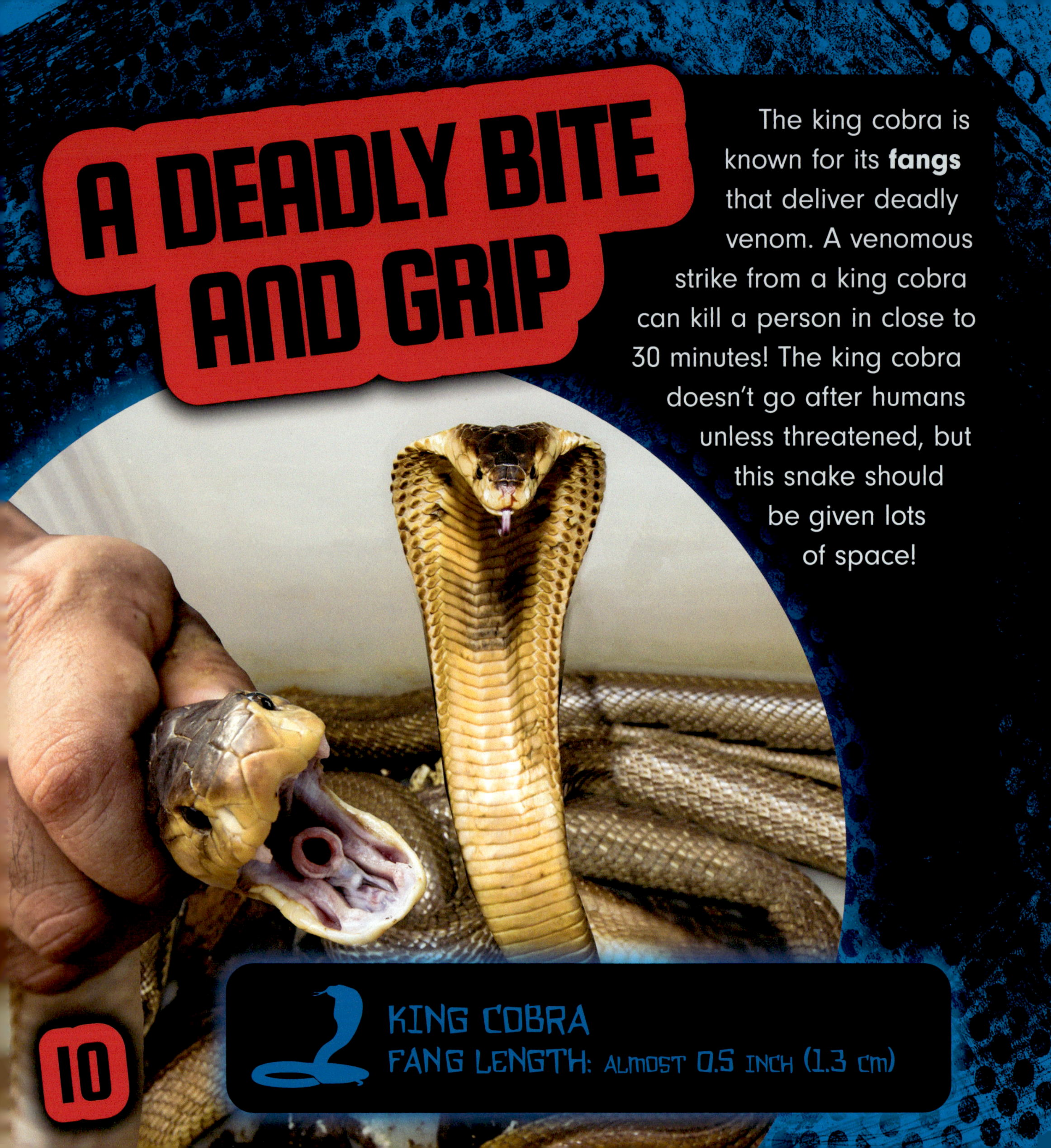

KING COBRA
FANG LENGTH: ALMOST 0.5 INCH (1.3 cm)

BALD EAGLE
TALON LENGTH: 2.1 INCHES (5.3 cm)

Bald eagles don't have venom like king cobras do, but that doesn't mean they're not tough fighters! Bald eagle beaks and **talons** are large and strong. A king cobra's venom is deadly, but the bald eagle's longest talons are longer than the cobra's fangs!

THE HUNTER OR THE HUNTED?

The king cobra might have deadly venom, but that doesn't mean they're the toughest predator out there! The king cobra is regularly eaten by predators that are **immune** to its venom or are too fast for the cobra to hit.

KING COBRA

NATURAL PREDATORS: HONEY BADGER, MONGOOSE, AND SECRETARY BIRD

BALD EAGLE
NATURAL PREDATORS: NONE

The bald eagle is considered an apex predator. This means that the bald eagle is at the top of its food chain and it has no natural predators! Unlike the king cobra, the only threat to the bald eagle is humans.

SHARP EYES

In a tough battle between predators, who can see the farthest is an important fact. King cobras have powerful eyesight that helps them see **prey** and predators from far away. This helps them spot their meal of choice: other snakes!

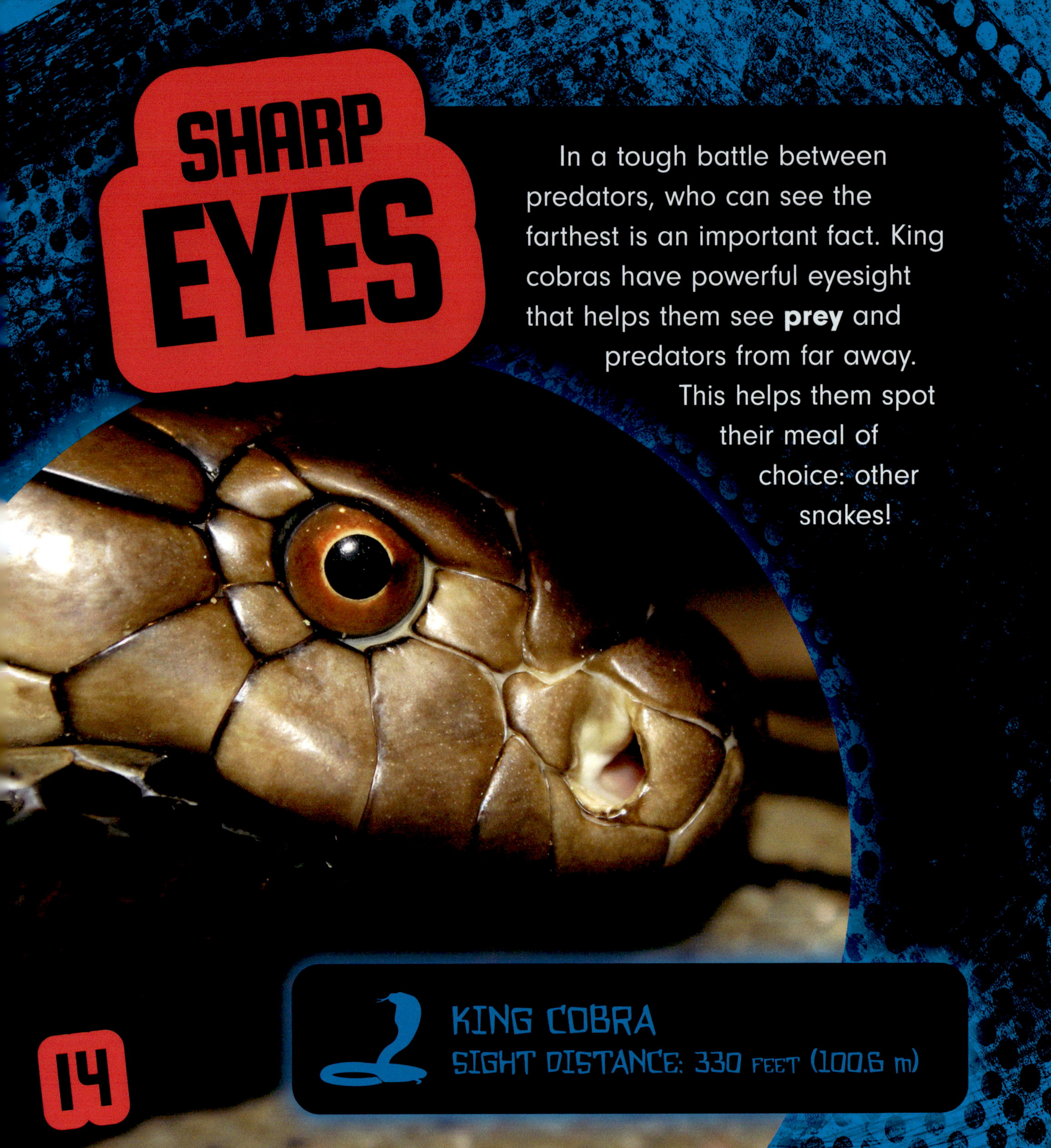

KING COBRA
SIGHT DISTANCE: 330 FEET (100.6 m)

The king cobra can see far, but the bald eagle can see farther. When an eagle is flying high up in the air, it can spot a rabbit moving on the ground. When they spot their prey, bald eagles can dive over 100 miles (161 km) per hour!

TINY BABIES

While the king cobra and the bald eagle both lay eggs, which predator lays the most? Between January and April, the female cobra lays her eggs and guards her nest. Female king cobras can be **aggressive** when keeping their nest safe.

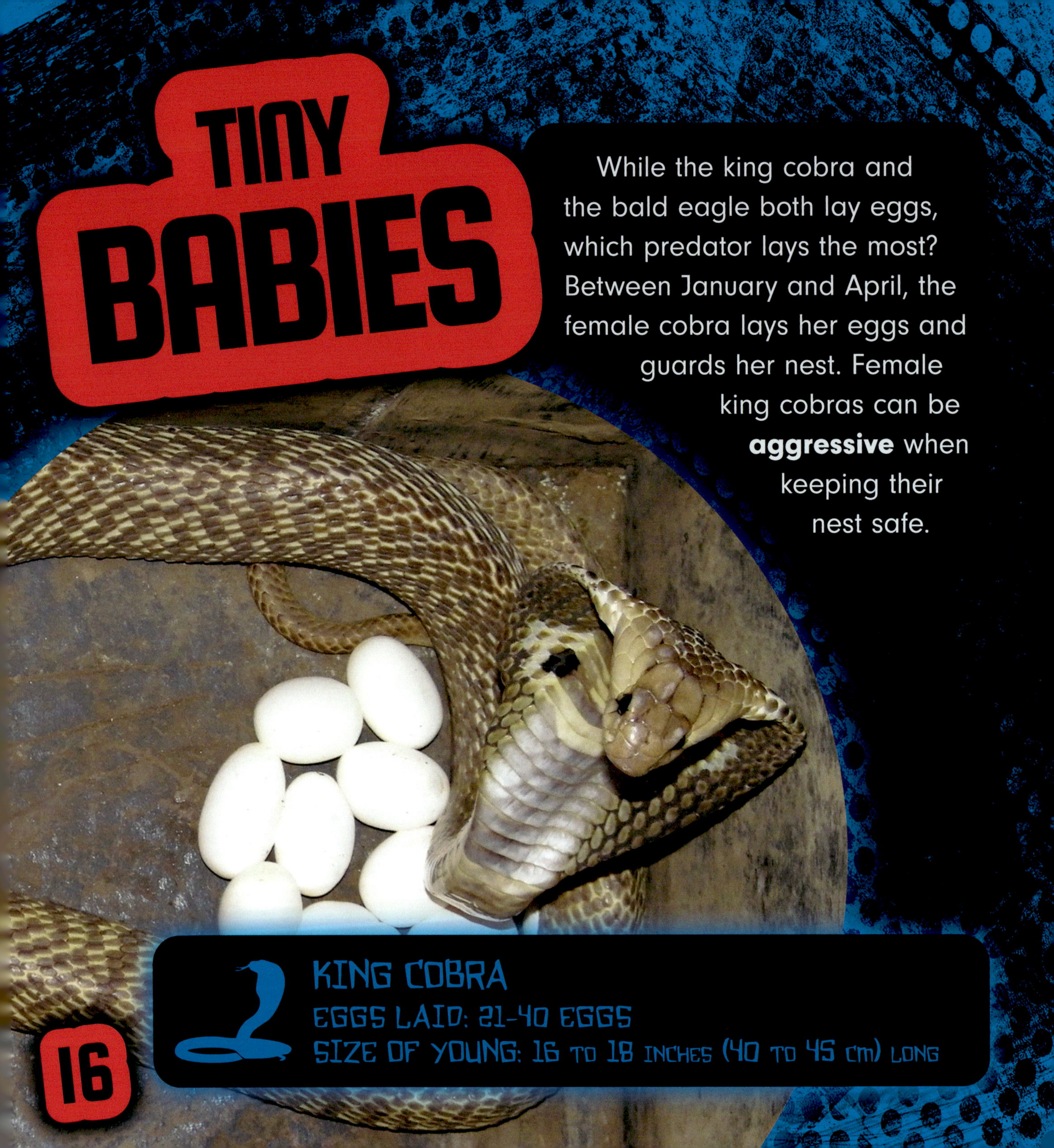

KING COBRA
EGGS LAID: 21-40 EGGS
SIZE OF YOUNG: 16 TO 18 INCHES (40 TO 45 CM) LONG

BALD EAGLE

NUMBER OF EGGS: 1 TO 3 EGGS

SIZE OF YOUNG: 2 OUNCES (60 G)

Bald eagles **mate** for life. This means that the same male and female eagle will remain together their whole lives. The king cobra may lay more eggs than the bald eagle, but the bald eagle will remain with their young for 18 weeks after the eaglets hatch, or break out from their shells!

THE LIFE OF A PREDATOR

Which predator lives the longest in the wild? Not only is the king cobra hunted by other animals, humans will often kill king cobras as well! But king cobras would rather stay far from humans that hurt them and usually end up near humans by mistake.

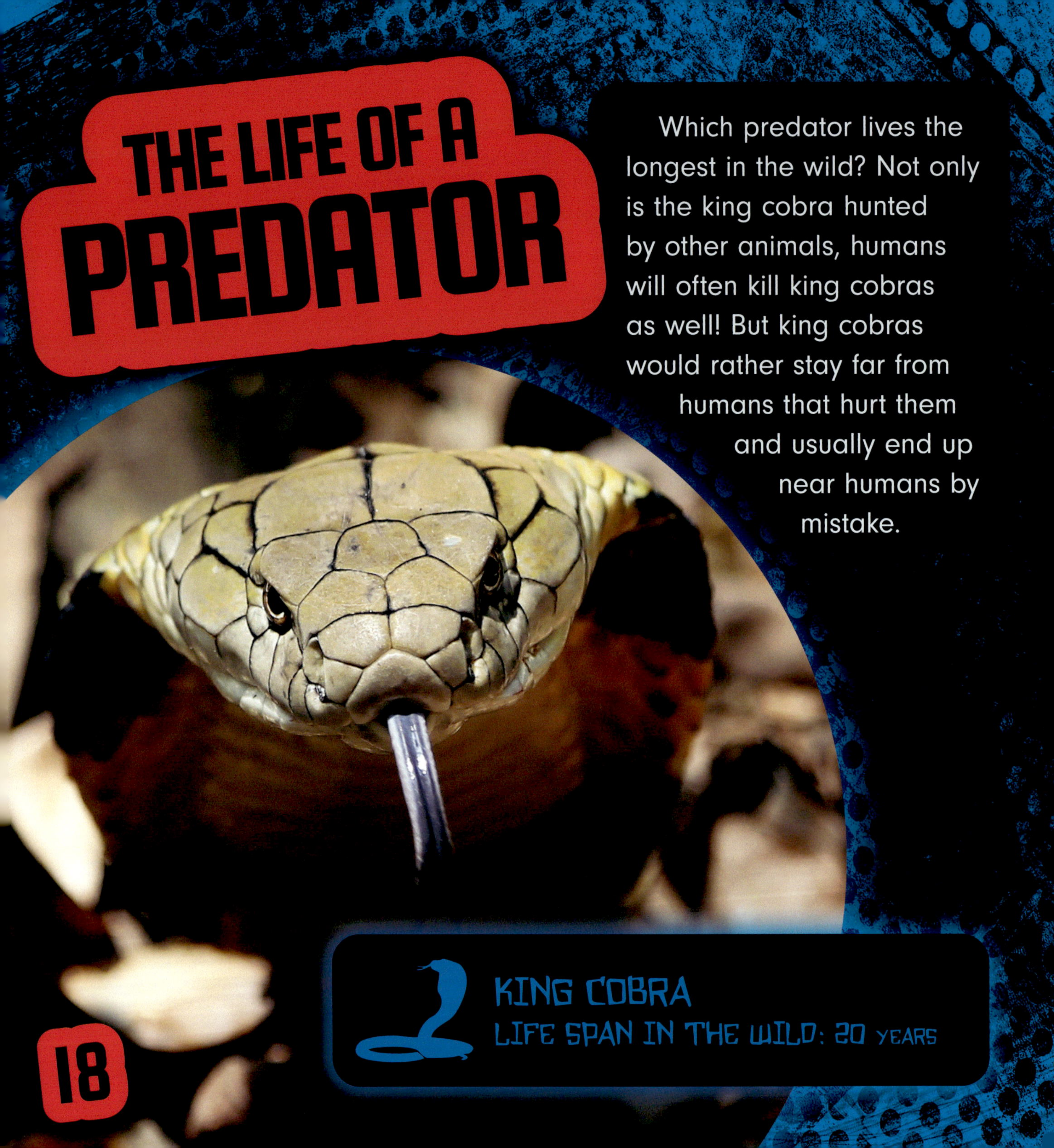

KING COBRA
LIFE SPAN IN THE WILD: 20 YEARS

BALD EAGLE

LIFE SPAN IN THE WILD: 15 TO 20 YEARS

When left alone, the bald eagle's life span is similar to the king cobra's. The biggest threat to the bald eagle's life span is humans. Because of pollution and losing their natural habitat, or home, the bald eagle population was threatened. **Conservation** efforts have helped bald eagles live longer and increase wild populations!

WHO WOULD WIN?

Who would win this battle? Would the bald eagle's speed in the air catch the king cobra off guard or would the king cobra's deadly venom get the eagle first? Would the king cobra's huge length be enough to scare off the bald eagle? Or would the eagle use its sharp talons to beat the cobra just like a secretary bird would?

In this beast battle, both fighters are too strong to back down! Who do you think would win?

BOTH OF THESE PREDATORS ARE TOP CONTENDERS, OR FIGHTERS, BUT IS THERE A CLEAR WINNER?

GLOSSARY

aggressive: Showing a readiness to attack.

conservation: The care of the natural world.

fang: A long, pointed tooth.

immune: To have protection from something.

mate: One of two animals that come together to produce babies.

prey: An animal that is hunted by other animals for food.

raptor: A bird of prey that kills other animals for food.

slither: To slide easily over the ground.

symbol: A picture, shape, or object that stands for something else.

talon: One of a bird's sharp claws.

threat: Something likely to cause harm. Threatened means likely to become endangered, or in danger of dying out.

venom: Matter an animal makes in its body that can harm other animals.

FOR MORE INFORMATION

BOOKS

Clasky, Leonard. *The Bald Eagle*. New York, NY: Gareth Stevens Publishing, 2023.

Jaycox, Jaclyn. *King Cobras*. North Mankato, MN: Pebble, 2023.

WEBSITES

National Geographic Kids: Bald Eagle
kids.nationalgeographic.com/animals/birds/facts/bald-eagle
Learn more about the bald eagle, where it lives, and what makes it an amazing bird!

National Geographic Kids: King Cobra
kids.nationalgeographic.com/animals/reptiles/facts/king-cobra
Check out more interesting facts about the king cobra and what makes it a tough predator.

Publisher's note to educators and parents: Our editors have carefully reviewed these websites to ensure that they are suitable for students. Many websites change frequently, however, and we cannot guarantee that a site's future contents will continue to meet our high standards of quality and educational value. Be advised that students should be closely supervised whenever they access the internet.

INDEX